# HUNTING LEGENDS

John Allan

## Picture Credits

(abbreviations: t = top; b = bottom; m = middle; l = left; r = right; bg = background)

**Shutterstock:** Catchlight Lens 7tl; Colin Edwards Wildside 14-15bg; DenisaPro 1bg, 18-19bg; Edwin Butter 24-25bg; Eric Isselee 4bg, 5bg; FloridaStock 28tr; Gary C. Tognoni 6tr; GUDKOV ANDREY 30ml; Imran Ashraf 3b, 8-9bg; Jane Rix 29tr; Jayaprasanna T.L 16-17bg; JayPierstorff 2tl, 7mr; Kshitij30 22-23bg; Johan Barnard 6bl; KGrif 31bl; Krasula 10-11bg; Ondrej Chvatal 20-21bg; Los t 31br; Milan Zygmunt 29br; SteffenTravel 26-27bg; Sourabh Bharti Howard 28br; Stu Porter 31tl; Ronnie Vaclav Sebek 30br.

Every effort has been made to trace the copyright holders and we apologise in advance for any unintentional omissions. We would be pleased to insert the appropriate credit in any subsequent edition of this publication.

Copyright © 2025 Hungry Tomato Ltd

First published in 2025 by Hungry Tomato Ltd
F15, Old Bakery Studios, Blewetts Wharf, Malpas Road, Truro, Cornwall,
TR1 1QH, UK.

No part of this publication may be reproduced, stored in a retrieval system, or transmitted in any form or by any means, electronic, mechanical, photocopying, recording, or otherwise, without prior written permission of the copyright owner.
A CIP catalogue record for this book is available from the British Library.

ISBN 9781835690789

Printed in China

Discover more at
www.hungrytomato.com

# HUNTING LEGENDS

## MEET SOME OF THE WORLD'S MOST DANGEROUS PREDATORS!

# CONTENTS

| | | | |
|---|---|---|---|
| Hunting Legends | 6 | Leopard | 20 |
| Great Horned Owl | 8 | Royal Bengal Tiger | 22 |
| Gila Monster | 10 | Polar Bear | 24 |
| Coyote | 12 | African Lion | 26 |
| Bald Eagle | 14 | When Predators Become Prey | 28 |
| Cheetah | 16 | Fearsome Facts | 30 |
| Wolverine | 18 | Glossary & Index | 32 |

Words in **BOLD** can be found in the glossary.

# HUNTING LEGENDS

Which animal wins the title of the deadliest **predator**? This big question isn't as easy to answer as you might think! There's lots to consider...

## THE DEADLIEST OF PREDATORS

We have captured some of the most dangerous and deadly predators on Earth within this book...

## PREDATORS BIG AND SMALL

The deadliest animals come in all shapes and sizes. Explore the top ten deadly all-stars, from the fierce African lion to the small but powerful wolverine...

## ON THE HUNT
Carnivores are animals that only eat meat, herbivores are animals that only eat plants, and omnivores are animals that eat both! Most of the extreme predators we explore in this book are carnivores, which makes them the most dangerous of all!

## DANGER COUNTDOWN
Each animal in this book is ranked in order, from the 10th deadliest creature, down to the ultimate predator. Which animal will come out on top?

# WARNING
## THINGS GET GRIM FROM HERE ON IN... TURN THE PAGES TO FIND OUT MORE!

# GREAT HORNED OWL

The great horned owl is an unusually deadly and dangerous predator, with the ability to **kill prey** up to three times its own size! It's even earned the nickname 'the tiger of the woods'.

## SUPER SKILLED
Great horned owls can't move their eyes like humans can. Instead, they must move their heads. These owls can almost completely turn their neck around! This is a massive advantage when it comes to looking for prey.

## STEALTHY HUNTERS

They mostly hunt at night, relying on their excellent sight and hearing. They dive down from high trees to snatch prey on the ground, flying and attacking without making any noise!

## FACT FILE

**WEIGHT**
Up to 2 kg (4 lbs)

**DIET**
Carnivore

**LOCATION**
Africa

**LETHAL POWERS**
Silent in flight, amazing senses, and **camouflaged** feathers

## DEADLY COUNTDOWN

# NO.10

# GILA MONSTER

This strange looking **reptile** is one of the few **venomous** lizards in the world. It has a powerful bite, and teeth covered in **venom** strong enough to kill an adult human!

### GREEDY GILA
Gila monsters can eat up to a third of their body weight in one meal! They catch small **mammals**, frogs and insects, and even eat bird eggs whole.

# FACT FILE

**WEIGHT**
Up to 2 kg (5 lbs)

**DIET**
Carnivore

**LOCATION**
North America

**LETHAL POWERS**
Powerful jaws, deadly venom, and strong teeth

## DEADLY COUNTDOWN

# NO.9

## A FRIGHTENING BITE
Once Gila monsters have bitten their prey, they use their strong jaws to hold on tight while the venom sets in. Some Gila monsters start chewing even when their prey is still alive!

# COYOTE

A coyote is a type of wild dog, that has **adapted** to lots of different environments. No matter if it's snowy landscapes or desert plains, coyotes can thrive! They are ruthless predators that hunt day and night.

## LONE HUNTERS
The coyote hunts alone, chasing prey over long distances without getting tired. They also have a great sense of smell that helps them sniff out prey hiding underground.

## SHARP WEAPONS
Coyotes have sharp claws, but their teeth are their main weapons. They have 42 in total, all very strong and very sharp!

# FACT FILE

**WEIGHT**
Up to 23 kg
(50 lbs)

**DIET**
Carnivore

**LOCATION**
North America

**LETHAL POWERS**
Lots of sharp teeth,
excellent sense of smell,
and high **stamina**

# DEADLY COUNTDOWN

# NO.8

# BALD EAGLE

The bald eagle is a ruthless hunter of the skies. Their wingspan can reach up to 2.5 metres (8 ft) long! They love to eat fish, often swooping down and using their sharp talons to grab fish straight out of the water!

## DEATH DROP

When bald eagles attack their prey, they can drop out of the sky at speeds of up to 100 mph (160 km/h)! They glide just above the water, using their impressive wing span. These birds don't hang around when it comes to hunting prey!

# FACT FILE

**WEIGHT**
Up to 6 kg
(14 lbs)

**DIET**
Carnivore

**LOCATION**
North America

**LETHAL POWERS**
Sharp talons, excellent eyesight, and speed

# DEADLY COUNTDOWN

# NO.7

## EAGLE EYED
Bald eagles have fantastic eyesight. They can spot a fish from 1 mile (1.6 km) away! Their eyes are around eight times stronger than ours!

# CHEETAH

The cheetah can run quicker than any other animal on Earth! It may be one of the smallest of the big cats, but it's also one of the deadliest. The cheetah uses its speed to catch fast-running prey on the grasslands of Africa.

## SNEAKY HUNTERS

Cheetahs hunt by creeping through long grass to get close to prey, before launching into a high-speed attack! These big cats attack with their teeth and claws, all at once.

## SPEEDY CATS

The cheetah has an unusually slim body, built for speed. Over short distances, it can reach a speed of nearly 60 mph (97 km/h).

# FACT FILE

**WEIGHT**
Up to 64 kg (140 lbs)

**DIET**
Carnivore

**LOCATION**
Africa

**LETHAL POWERS**
Super fast, sharp teeth, and stealthy hunters

# DEADLY COUNTDOWN NO.6

# WOLVERINE

A wolverine might just look like a small bear, but it is actually the largest and fiercest member of the weasel family. It chases after small prey, but is known to climb trees to attack larger prey from above!

## HUNGRY HUNTERS

When following prey, wolverines can travel for up to 40 miles (64 km) a day. They can be very aggressive hunters and have been known to fight animals much bigger than themselves, like wolves and small bears, over food.

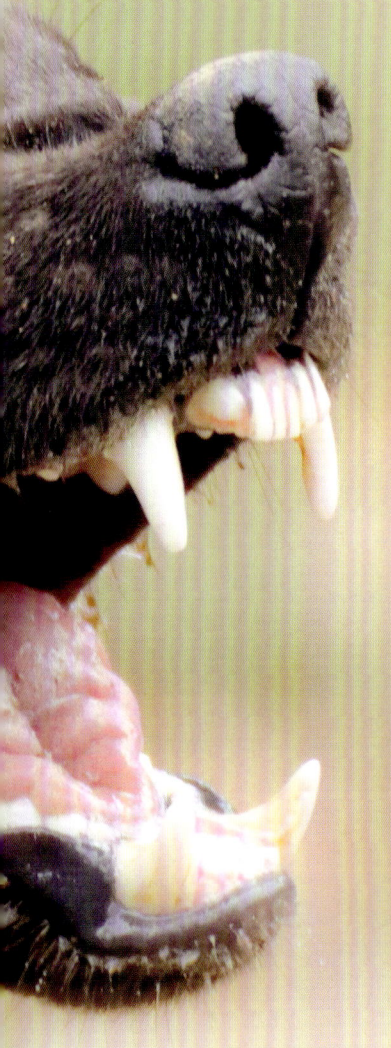

## JAW-DROPPING

Wolverines have very strong jaws - they can bite through frozen meat, and even bone!

# FACT FILE

### WEIGHT
Up to 18 kg (40 lbs)

### DIET
Omnivore

### LOCATION
Asia, Europe, and North America

### LETHAL POWERS
Powerful jaws, high stamina, and agressive

# DEADLY COUNTDOWN

# NO.5

# LEOPARD

The leopard is a secretive and deadly predator that hunts for prey at night! Powerful legs and neck muscles make it the strongest climber of the big cats. It can sneak up on its prey through tall grass or jump down on them from a tree branch above!

## STRONG HUNTERS
Unlike cheetahs who rely on speed for hunting, these big cats rely on strength. Leopards will often hunt and kill prey much larger than themselves!

## CLEVER CAMOUFLAGE
Their spotted fur is perfect for camouflage, and their large paws contain very sharp claws for gripping onto prey.

# FACT FILE

### WEIGHT
Up to 75 kg (165 lbs)

### DIET
Carnivore

### LOCATION
Asia and Africa

### LETHAL POWERS
Powerful limbs, excellent camouflage, and sharp, curved claws

## DEADLY COUNTDOWN
# NO.4

# ROYAL BENGAL TIGER

The size and power of the Royal Bengal tiger makes it an incredible predator. It is capable of killing and carrying prey much bigger than itself! It's prey include antelope, wild boar, and water buffalo, and it has even been known to develop an appetite for humans...

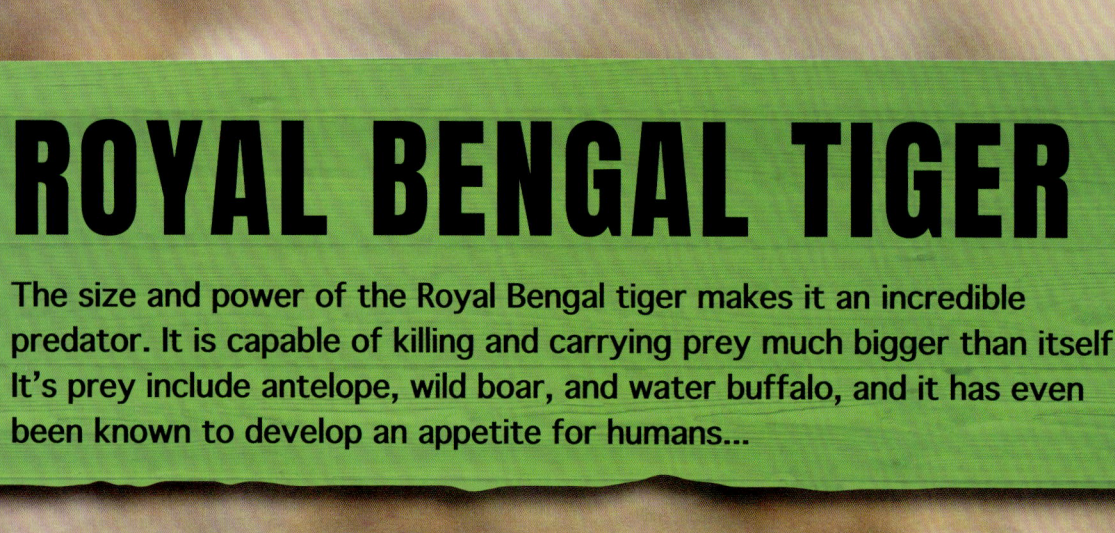

## LETHAL HUNTERS
Royal Bengal tigers hunt by leaping onto their prey, dragging them down with sharp claws, and biting their throats to kill them. They have the biggest teeth among the big cats, using them to take down animals larger than themselves.

## EXCELLENT EYES
They have excellent vision, seeing six times better than humans!

## FACT FILE

**WEIGHT**
Up to 225 kg (500 lbs)

**DIET**
Carnivore

**LOCATION**
Indian subcontinent

**LETHAL POWERS**
Excellent vision, fast runners, and large sharp teeth

## DEADLY COUNTDOWN NO.3

23

# POLAR BEAR

Polar bears are the largest and most powerful predators that live on land, and they have nothing to fear – except hunger! During warmer months, these huge creatures often go hungry because they can't hunt seals unless the sea is frozen. In these difficult times, a polar bear might not eat for months.

## PATIENT HUNTERS
Polar bears will catch seals by waiting for them to come up to the surface. They use their sharp claws to easily rip through skin and muscle, and their tough jaws to crunch through bone.

**SUPER SWIMMERS**
Polar bears can swim constantly for days at a time, searching for prey!

# FACT FILE

**WEIGHT**
Up to 770 kg
(1,700 lbs)

**DIET**
Carnivore

**LOCATION**
Arctic

**LETHAL POWERS**
Powerful muscles, strong jaws, and excellent swimmer

# DEADLY COUNTDOWN

# NO.2

# AFRICAN LION

This lion is Africa's top predator, and for good reason! The females often communicate and hunt together to take down large animals, leaving the males to defend their **territory.** Lions are best known for their almighty roar!

## FIERCE FEATURES

African lions have five claws on each paw, perfect for grabbing onto and killing prey. Their powerful jaws are packed with 30 sharp teeth that easily rip through flesh.

## HERD HUNTING

When lions hunt, they help each other to make sure the prey is surrounded at all times so it can't escape. That's deadly teamwork!

# FACT FILE

### WEIGHT
Up to 190 kg (420 lbs)

### DIET
Carnivore

### LOCATION
Africa

### LETHAL POWERS
Hunt in groups, five claws per paw, and sharp teeth

## DEADLY COUNTDOWN

# NO.1

# WHEN PREDATORS BECOME PREY

These predators are some of the deadliest in the animal kingdom. But what happens when these impressive hunters feel like prey themselves?

## POLAR BEAR

Polar bears are apex predators – this means they're at the top of the food chain! However, climate change is causing their icy hunting grounds to melt. This loss of habitat means they often go hungry and their numbers are shrinking.

## COYOTE

While excellent hunters, coyotes sometimes find themselves at the other end of the dinner table! Bears, wolves, and birds of prey hunt these wild dogs.

## CHEETAH

Although cheetahs are the fastest land animal alive, they still face threats! As animal numbers decline, there is less prey for cheetahs to chase, meaning this speedy **species** will often go hungry.

## ROYAL BENGAL TIGER

Royal Bengal tigers are dying out because of human activity, such as illegal hunting for tiger skin. There are less than 2,500 of these tigers left in the wild!

# FEARSOME FACTS

There are so many deadly and fearsome facts about these incredible predators. Here are some more that show just how impressive each of these animals are...

### POLAR BEAR
Male polar bears are twice the size of females and can weigh the same as ten humans!

### COYOTE
Coyotes use howling, yapping, and leaving their scent as ways to mark their territory, and to help identify others coyote groups.

### GILA MONSTER
Gila monsters spend most of their time underground, as they don't need to hunt very often. They store extra fat in their tails to give them energy.

### CHEETAH
The black markings on a cheetah's face help with their eyesight! In the bright sun, these markings redirect the sunlight off their faces and help them to see.

## AFRICAN LION
Young African lions are born with spots on their fur, but these often go away after they become adults.

## GREAT HORNED OWL
This owl doesn't really have horns; it has feathers that sit on the top of its head and look like horns!

## LEOPARD
Leopards are solitary creatures, which means they like to live alone. Each leopard has their own territory, which they mark with their scent, poop, and pee!

## ROYAL BENGAL TIGER
Despite their size, Royal Bengal tigers are excellent swimmers! They are known to chase prey to the water's edge to make them easier to catch.

## BALD EAGLE
Bald eagles create the largest bird nests! They can be up to 2 metres (6 feet) wide, and are often found at the top of trees.

## WOLVERINE
Wolverines love to climb! They don't chase or creep up on their prey. They prefer to pounce on them from above!

# GLOSSARY

**Adapted** - able to fit in with surroundings.

**Camouflage** – when an animal blends into its surroundings so it's hard to see.

**Herd** – a group of animals of one kind that live or travel together.

**Mammals** – warm-blooded animals with a covering of hair on the skin and the ability to produce milk to feed their young.

**Predator** – an animal that lives by attacking and killing other animals.

**Prey** – an animal hunted or caught for food.

**Reptile** – a cold-blooded animal that has scales and lays eggs on land.

**Species** – a group of living things that have the same features as each other and share a common name.

**Stamina** – lasting strength and energy.

**Territory** - a specific area that belongs to or is controlled by someone or something.

**Venom** – a poisonous substance of an animal, usually passed on by a bite or a sting.

**Venomous** – a creature that can produce venom (see above).

# INDEX

**A**
African lion 6, 26-27, 31

**B**
Bald eagle 14-15, 31

**C**
Carnivore 7, 9, 11, 13, 15, 17, 21, 23, 25, 27
Cheetah 16-17, 29, 30
Coyote 12-13, 28, 30

**G**
Gila monster 10-11, 30
Great horned owl 8-9, 31

**H**
herbivore (plant-eater) 7

**L**
Leopard 20-21, 31

**O**
omnivore 7, 19

**P**
Polar bear 24-25, 28, 30

**R**
Royal Bengal tiger 22-23, 29, 31

**W**
Wolverine 6, 18-19, 31